YOUR KNOWLEDGE HAS VALUE

- We will publish your bachelor's and
  master's thesis, essays and papers

- Your own eBook and book -
  sold worldwide in all relevant shops

- Earn money with each sale

Upload your text at www.GRIN.com
and publish for free

Imprint:

Copyright © 2016 GRIN Verlag
Print and binding: Books on Demand GmbH, Norderstedt Germany
ISBN: 9783668688988

Suresh Kumar Venugopal, Nithya Ranganathan, Narendhar Chandrasekar

Green Synthesis of Silver Nanoparticles Usim Triticum Aestivum and Emblica Officinalis. Characterization and their Antibacterial Study

GRIN Verlag

Contents

INTRODUCTION
NANOTECHNOLOGY

Nanotechnology is emerging as a rapidly growing field with its application in science and technology for the purpose of manufacturing new materials at the nanoscale level (Albrecht *et al., 2006*). Nanotechnology is a field of science which deals with production, manipulation and use of materials ranging in nanometers.

Nanotechnology ("nanotech") is manipulation of matter on an atomic, molecular, and supramolecular scale. The earliest, widespread description of nanotechnology referred to the particular technological goal of precisely manipulating atoms and molecules for fabrication of macroscale products, also now referred to as molecular nanotechnology. (Drexler *et al.,* (1986)

A more generalized description of nanotechnology was subsequently established by the National Nanotechnology Initiative, which defines nanotechnology as the manipulation of matter with at least one dimension sized from 1 to 100 nanometers. This definition reflects the fact that quantum mechanical effects are important at this quantum-realm scale, and so the definition shifted from a particular technological goal to a research category inclusive of all types of research and technologies that deal with the special properties of matter that occur below the given size threshold. It is therefore common to see the plural form "nanotechnologies" as well as "nanoscale technologies" to refer to the broad range of research and applications whose common trait is size.(Drexler *et al.,* (1992)

Nanotechnology as defined by size is naturally very broad, including fields of science as diverse as surface science, organic chemistry, molecular biology, semiconductor physics, microfabrication, etc. The associated research and applications are equally diverse, ranging from extensions of conventional device physics to completely new approaches based upon molecular self-assembly, from developing new materials with dimensions on the nanoscale to direct control of matter on the atomic scale.

NANOBIOTECHNOLOGY

A new branch of nanotechnology is nanobiotechnology. Nanobiotechnology combines biological principles with physical and chemical procedures to generate nano-sized particles with specific functions. Nanobiotechnology represents an economic alternative for chemical and physical methods of nanoparticles formation (Ahmad *et al.,* 2003a). Nanobiotechnology is defined as a field that applies the nanoscale principles and techniques to understand and transform biosystems (living or non-living) and which uses biological principles materials to create a new devices and systems integrated from the nanoscale. The integration of nanotechnology with biotechnology, as well as with infotechnology and cognitive science, is expected to accelerate in the next decade (Roco and Bainbridge, 2002).

Bionanotechnology and nanobiotechnology are terms that refer to the intersection of nanotechnology and biology (Ehud Gazi, 2007). These two terms are often used interchangeably. Bionanotechnology generally refers to the study of how the goals of nanotechnology can be guided by studying how biological "machines" work and adapting these biological motifs into improving existing nanotechnologies or creating new ones (Nolting,2005).

Nanobiotechnology takes most of its fundamental from nanotechnology. Most of the devices designed for nanobiotechnological use are directly based on other existing nanotechnologies. Nanobiotechnology is often used to describe the overlapping multidisciplinary activities associated with biosensors, particularly where photonics, chemistry, biology, biophysics, nanomedicine, and engineering converge. Nanobiotechnology is extending the limits of current molecular diagnostics and is facilating point of care diagnosis, Integration of diagnostics with therapeutics and thus advancing the development of personalized medicine (Jain,2007).

Nanobiotechnology is a burgeoning interdisciplinary field of research interlacing material science, bionanoscience and technology. The advances made in the field of nanobiotechnology to harness the benefit of life sciences, health care and industrial biotechnology are remarkable. Based on these aspects metal nanoparticles have received considerable attention in recent years because of their potential applications in biological tagging and pharmaceutical applications, which have a high specific surface area and fraction of surface atoms, though the nanoparticles like copper, zinc oxide, gold, alginate and silver have proved to be most effective antimicrobial activity against numerous bacteria and viruses. (Kumaravel palanisamy *et al.*, (2014)).

NANOPARTICLES

Nanoparticles can be easily synthesized using various methods by various approaches available for the synthesis of silver nanoparticles include chemical, electrochemical, radiation, photochemical methods and Langmuir-Blodgett and biological techniques. But most of the chemical methods used for the synthesis of nanoparticles involve the use of toxic, hazardous chemicals that create biological risks and sometime these chemical processes are not ecofriendly. This enhances the growing need to develop environmentally friendly processes through green synthesis and other biological approaches.

Metal nanoparticles were widely used in different application with the rapid growth of nanoparticles (AgNPs) widely used in antimicrobial activity, cosmetics industry and daily products. Synthesis method of nanoparticles plays an important role in its technological advancement as it is the primary step to tune their physical, electronic and optical properties of the synthesized nanoparticles by varying size, shape and surface chemistry (Sharma *et al.*, 2009).

Metal nanoparticles are now studied extensively because of their unique physicochemical characteristics including catalytic activity, optical property, electronic property, magnetic property and antimicrobial activity. Synthesis of noble nanoparticles for applications in these areas is of current research interest. Generally, metal nanoparticles are synthesized by using chemical routes such as chemical reduction, phytochemical reactions in reverse micelles and recently using routes of green chemistry.

The medicinal and preservative properties of silver have been known for over 2000 years. Silver based compounds have been widely used in bactericidal applications, in burn, wound therapy and healthcare product (Klasen 2000). They have been widely used as disinfectant that inhibit bacteria growth by inhibiting the essential enzymatic functions of the microorganism and attribution to reactive oxygen species mediated bactericidal activity (Bragg and Rainnie 1974; Messner and Imlay 1999). Currently silver nanoparticles are wildly used as antibacterial and antifungal agents in a diverse range of consumer products: air sanitizer sprays, detergents, soaps, shampoos, toothpastes and washing machine (Buzea *et al.* 2007). Nanoparticles exhibit new or improved properties depending upon their size, morphology and distribution (Song and Kiml 2009; Nalwa 2000).

SYNTHESIS METHODS
Physical methods

Evaporation-condensation and laser ablation are the most important physical approaches. The absence of solvent contamination in the prepared thin films and the uniformity of NPs distribution are the advantages of physical synthesis methods in comparison with chemical processes. (Kruis *et al.*, 2000).

Silver NPs could be synthesized by laser ablation of metallic bulk materials in solution (Mafune *et al.*, 2000). The ablation efficiency and the characteristics of produced nano-silver particles depend upon many parameters, including the wavelength of the laser impinging the metallic target, the duration of the laser pulses (in the femto-, pico- and nanosecond regime), the laser fluence, the ablation time duration and the effective liquid medium, with or without the presence of surfactants (Kim *et al.*, 2005).

Chemical Methods

The most common approach for synthesis of silver NPs is chemical reduction by organic and inorganic reducing agents. In general, different reducing agents such as sodium citrate, ascorbate, sodium borohydride (NaBH4), elemental hydrogen, polyol process, Tollens reagent, N, N-dimethylformamide (DMF), and poly (ethylene glycol)-block copolymers are used for reduction of silver ions (Ag+) in aqueous or non-aqueous solutions. These reducing agents reduce Ag+ and lead to the formation of metallic silver (Ag0), which is followed by agglomeration into oligomeric clusters. These clusters eventually lead to the formation of metallic colloidal silver particles (Wiley *et al.*, 2005). It is important to use protective agents to stabilize dispersive NPs during the course of metal nanoparticle preparation, and protect the NPs that can be absorbed on or bind onto nanoparticle surfaces, avoiding their agglomeration (Oliveira *et al.*, 2005). The presence of surfactants comprising functionalities (e.g., thiols, amines, acids, and alcohols) for interactions with particle surfaces can stabilize particle growth, and protect particles from sedimentation, agglomeration, or losing their surface properties.

Polymeric compounds such as poly (vinyl alcohol), poly (vinylpyrrolidone), poly (ethylene glycol), poly (methacrylic acid), and polymethylmethacrylate have been reported to be the effective protective agents to stabilize NPs (Brust *et al.*, 2005).

Microemulsion techniques

Uniform and size controllable silver NPs can be synthesized using microemulsion techniques. The NPs preparation in two-phase aqueous organic systems is based on the initial spatial separation of reactants (metal precursor and reducing agent) in two immiscible phases. (Krutyakov *et al.*, 2008)

UV-initiated photoreduction

A simple and effective method, UV-initiated photoreduction, has been reported for synthesis of silver NPs in the presence of citrate, polyvinylpyrrolidone, poly (acrylic acid), and collagen. (Huang and Yang,2008)

NEED FOR BIOLOGICAL METHODS

The need for biosynthesis of nanoparticles rose as the physical and chemical processes were costly. So in the search of for cheaper pathways for nanoparticle synthesis, scientists used microorganisms and then placed extracts for synthesis. Nature has devised various processes for the synthesis of nano- and micro- length scaled inorganic materials which have contributed to the development of relatively new and largely unexplored area of research based on the biosynthesis of nanomaterials (Mohanpuria *et al.*, 2008). Use of fungi, bacteria and plant extracts for synthesis of nanoparticles is quite novel leading to green chemistry which provides advantages over chemical and physical methods as it is cost-effective and environment-friendly, and can be scaled up for large scale preparation. (Kaushik Roy *et al.*, 2013).

Bioinspired synthesis of nanoparticles provides advancement over chemical and physical methods as it is a cost effective and environment friendly and in this method there is no need to use high pressure, energy, temperature, and toxic chemicals (Goodshell 2004).

Biosynthesis of nanoparticles is a kind of bottom up approach where the main reaction occurring is reduction/oxidation. The microbial enzymes or the plant phytochemicals with anti oxidant or reducing properties are usually responsible for reduction of metal compounds into their respective nanoparticles.

USES OF PLANTS FOR NANOPARTICLE SYNTHESIS

The advantage of using plants for the synthesis of nanoparticles is that they are easily available, safe to handle and posses a broad variability of metabolites that may aid in reduction. A number of plants are being currently investigated for their role in the synthesis of nanoparticles.

Sometimes the synthesis of nanoparticles using various plants materials and their extracts can be beneficial over other biological synthesis processes which involve the very complex procedures of maintaining microbial cultures. (Swarup Roy *et* al., 2015).

The use of plants as the production assembly of silver nanoparticles has drawn attention, because of its rapid, ecofriendly, non-pathogenic, economical protocol and providing a single step technique for the biosynthetic processes. The reduction and stabilization of silver ions by combination of biomolecules such as proteins, amino acids, enzymes, polysaccharides, alkaloids, tannins, phenolics, saponins, terpinoids and vitamins which are already established in the plant extracts having medicinal values and are environmental benign, yet chemically complex structures (Kulkarni *et al.*, 2014).

A vast segment of flora had been utilized for the preparation of silver nanoparticles. Different plants and their respective portions have been exploited for the same as well.

OBJECTIVE OF THE STUDY

✓ To extract the compounds *from Triticum aestivum & Emblica officinalis* for Silver Nanoparticle Synthesis.

✓ To synthesize Silver nanoparticles using plant extracts of *Triticum aestivum & Emblica officinalis*

✓ To characterize the synthesized silver nanoparticles using UV-VIS, SEM & DLS.

✓ To identify the compound responsible for nanoparticle synthesis using FTIR and anti-microbial action.

✓ To study the anti-microbial activity of the synthesized silver nanoparticles and comparison with crude extract.

REVIEW OF LITERATURE

Stephen *et al.*, in (1999) and Gardea Torresdey *et al.*, in (2002) has reported green chemistry approach that interconnects nanotechnology and biotechnology.

Naik *et al.*, (2002); Wiley *et al.*, (2007); Sharma *et al.*, (2009) has reported the synthesis methods of silver nanoparticles. Biosynthetic methods employing either biological microorganisms such as bacteria Joerger *et al.*, (2000) and fungus ;Shankar *et al.*, (2003a) or plants extract (Gardea-Torresdey *et al.*, 2002; Shankar *et al.*, (2003b); Chandran *et al.*, (2006;), have emerged as a simple and viable alternative to more complex chemical synthetic procedures to obtain nanomaterials. Different types of nanomaterials like copper, zinc, titanium

(Retchkiman-schabes *et al.*, 2006), magnesium, gold ;Gu *et al.*, (2003), alginate; Ahmad *et al.*, (2005) and silver have come up but silver nanopartic les have proved to be most effective as it has good antimicrobial efficacy against bacteria, viruses and other eukaryotic microorganisms; Gong *et al.*, (2007).

Plant extract, and enzymes have proved itself an easy, cost-effective and ecofriendly alternative synthesis route of metallic nanoparticle compared to conventional procedures (Li *et al.*, (2007); Mittal *et al.*, (2013); Tamulya *et al.*, (2013).

Numerous reports have already been established to synthesize metal nanoparticles using plant extract of *Aloevera* (Chandran *et al.*, (2006), *Phyllanthus emblica* (Saini *et al.*, (2013), *Azadirachta indica* (Saini *et al.*, (2013); Lalitha *et al.*, (2013), *Rhinacanthus nasutus* (Pasupuleti *et al.*, (2013), *Memecylon edule* (Elavazhagan and Arunachalam 2011), *Magnolia kobus* (Lee *et al.*, (2013), *Cymbopogan citrates* (Masurkar *et al.* (2011), *Morinda tinctoria* (Vanaja *et al.*, (2014), *Malus domestica* (Roy *et al.*, (2014).

In continuation of this efforts for synthesizing Ag NPs by green route, a facile, rapid, cheaper and single pot aqueous biosynthesis of Ag NPs using the leaves extract of *E.officinalis* have been reported by hepatoprotective; Jeena *et al.*, (1991), De *et al.*, (1993); Jeena *et al.*, (1995); Thakur *et al.*, (1998); Ahmad *et al.*, (1998), Bhattacharya *et al.*, (2000)), antitumor; Jose *et al.*, (2001), Sabu *et al.*, (2002); antiulcerogenic (Sri Ram *et al.*, (2002), AlRehaily *et al.*, (2002) anti-inflammatory; Sharma *et al.*, (2003), analgesic and antipyretic; Perianayaham *et al.*, (2004), and antidiarrheal; Perianayaham *et al.*, (2005).In contrast to this several groups have achieved success in the plant mediated biological synthesis of Ag NPs using extracts obtained from *Anacardium occidentale*; Sheny *et al.*, (2011), *Malva parviflora* ;Mervat *et al.*, (2012), *Ocimum tenuiflorum*; Rupali *et al.*, (2012), *Gloriosa superb* ;Ashokkumar *et al.*, (2013), *Hibiscus cannabinus*; Bindu *et al.*, (2013), *Sesbania grandiflora*; Das *et al.*, (2013), *Mangifera indica* ;Daizy Philip, (2011), *Prosopis juliflora* ; Raja *et al.*, (2012) and *Cocculus hirsutus;* Thiruppathi *et al.*, (2013).

The reduction of silver ions to silver nanoparticles by this extract was completed within 10 min. The extracellular silver nanoparticles syntheses by aqueous leaf extract validate quick, simple, economical process comparable to chemical and microbial methods. These silver nanoparticles exhibit antibacterial activity against *Pseudomonas aeruginosa*, *Escherichia coli*, *Klebsiella pneumonia* and *Enterococcus faecal* ; Kumar *et al.*, (2014). *Acorus calamus* was also used for the synthesis of silver nanoparticles to evaluate its antioxidant, antibacterial as well as anticancer effects ; Nakkala *et al.*, (2014). *Boerhaavia diffusa* plant extract was used as a reducing agent for green synthesis of silver nanoparticles.

Roy & Das (2015) With the advancement of technologies and improved scientific knowledge a way for research and development in the field of herbal and medicinal plant biology towards intersection of nanotechnology has been observed. One such interference is applying plants source in the green synthesis of nanoparticles.

Jose *et al.*, (2005) and Lok *et al.*, (2007) Silver has long been recognized as having inhibitory effect on microbes present in medical and industrial process.

Ip *et al.*, (2006) The most important application of silver and silver nanoparticles is in medical industry such as topical ointments to prevent infection against burn and open wounds.

Wheatgrass is a good source of potassium, a very good source of dietary fiber, vitamin A, vitamin C, vitamin E (alpha tocopherol), vitamin K, thiamin, riboflavin, niacin, vitamin B6, pantothenic acid, iron, zinc, copper, manganese and selenium, and has a negligible amount of protein (less than one gram per 28 grams). Melina et al., (2003); Meyrowitz, Steve (1999).

In Ayurvedic polyherbal formulations, Indian gooseberry is a common constituent, and most notably is the primary ingredient in an ancient herbal rasayana called Chyawanprash (Dharmananda, 2008). This formula, which contains 43 herbal ingredients as well as clarified butter, sesame oil, sugar cane juice, and honey, was first mentioned in the Charaka Samhita as a premier rejuvenative compound. (Samhita, 1949). In Chinese traditional therapy, this fruit is called yuganzi, which is used to treat throat inflammation.

WHEAT GRASS (<u>Triticum</u> <u>aestivum</u>)

<u>Triticum</u> <u>aestivum</u>

Kingdom	: <u>Plantae</u>
(unranked)	: <u>Angiosperms</u>
(unranked)	: <u>Monocots</u>
(unranked)	: <u>Commelinids</u>
Order	: <u>Poales</u>
Family	: <u>Poaceae</u>
Subfamily	: <u>Pooideae</u>
Tribe	: <u>Triticeae</u>
Genus	: *<u>Triticum</u>*
Species	: ***T. aestivum***

<u>Emblica officinalis</u>

Kingdom	: <u>Plantae</u>
(unranked)	: <u>Angiosperms</u>
(unranked)	: <u>Eudicots</u>
(unranked)	: <u>Rosids</u>
Order	: <u>Malpighiales</u>
Family	: <u>Phyllanthaceae</u>
Tribe	: <u>Phyllantheae</u>
Subtribe	: <u>Flueggeinae</u>
Genus	: *<u>Phyllanthus</u>*
Species	: ***P. emblica***

MATERIALS AND METHODS

Collection of Plant materials & Processing

Two different plants were selected for the silver nanoparticles synthesis. The Leaves of Wheat Grass (*Triticum aestivum*) and Fruit of Gooseberry (*Emblica officinalis*) were collected. The collected leaves were washed aseptically with distilled water, 0.1% Mercuric chloride and kept under shade for drying and the fruit sample were washed with distilled water. After shade drying the Wheat grass sample were powdered.

Extraction

Wheat Grass (*Triticum aestivum*) extract was prepared from the powder of the plant sample. 5g of plant powder was mixed with the 100ml of distilled water and boiled for 10 minutes. The plant extract was allowed to cool at room temperature and then filtered with Whatman filter paper No. 1 (25 μm pore size) and was stored in 4°c. (Natthawan Phuphansari *et al.*, 2012)

Goose Berry (*Emblica officinalis*) extract was prepared from the fruit sample. The collected fruits were washed with distilled water and dried with water absorbent paper. Then it was cut into small pieces with an ethanol sterile knife and crushed with mortor and pestle and dispensed in 10ml of sterile distilled water. Then extract was then filtered using Whatman filter paper No. 1 (25 μm pore size) and was stored in 4°c. (Natthawan Phuphansari *et al.*, 2012)

Synthesis of Silver Nanoparticles

Silver nitrate solution (50 ml, 1mM) and Wheat grass (*Triticum aestivum*) extract (2ml) were mixed and mixture was constant stirred and kept at room temperature to react.

Silver nitrate solution (50ml, 20mM) and Goose berry (*Emblica officinalis*) extract (2ml) were mixed and mixture was constant stirred and kept at room temperature to react.

Characterization

UV-Visible Spectroscopy analysis

To study the formation of Silver Nanoparticles, the UV-Visible (UV-Vis) spectrophotometer was used.

Fourier transformed infrared spectroscopy (FT-IR)

The synthesized silver nanoparticles were selected to FT-IR studies.

Dynamic Light Scattering (DLS)

Dynamic light scattering (or Photon correlation Spectroscopy) is an important technique generally used to realize the size distribution pattern of very small particles present in suspension or solution. This light scattering technique was used in this study to realize the size distribution profile of synthesized silver nanoparticles.

SEM Analysis

Scanning Electron Microscopy was used to investigate the microstructure and crystallinity of the silver nanoparticles synthesized using leaf extract of Wheat Grass (*Triticum aestivum*) and plant extract of Goose Berry (*Emblica officinalis*).

Antimicrobial Study

By the disc diffusion method, the antibacterial activities of the Wheat grass and Goose berry extract synthesized silver nanoparticles were studied. Muller Hinton agar plate was prepared, sterilized and solidified. Then two bacterial Pathogens (*Klebsiella*, *Staphylococcus aureus*) were swabbed on the plates. Sterile discs were dipped in silver nanoparticles solution (30 μl and 50 μl) and placed in the Muller Hinton Agar plate and kept at incubation for 37°C for 24 hours. Silver nanoparticles, antibiotic disc and crude extract were maintained as positive control. After incubation the Zone of Inhibition was measured.

RESULTS

Synthesis of Silver Nanoparticles

The synthesis of silver nanoparticles using a mixture of AgNO3 (1mM and 20mM, 50ml) and 2ml of extract at room temperature for 1 h was preliminary studied. The change of color was observed. The solution color changed to brown shade from yellow wheat grass extract and the colourless solution of goose berry extract was to dark brown color in 24h. (Fig.1)

UV-Visible Spectroscopy

The progress of reaction of synthesized silver nanoparticles was studied by UV-visible spectroscopy. The peak intensity related to the concentration of silver when the reaction time extended. The absorption band of Goose berry is about 210 nm and Wheat grass is about 399.5 nm. (Fig.2 and Fig.3)

Fourier transforms infrared spectroscopy (FT-IR)

Silver nanoparticles synthesized using Wheat grass extract was used for FT-IR analysis. FT-IR spectroscopy measurements are carried out to identify the biomolecules that bound specifically on silver surface and the molecular environment of the capping agent on the nanoparticles. The spectrum of silver nanoparticles extract of wheat grass showed an intense peak at 3321.20 cm⁻1. The compound was found to be Alkyne Group denotes the C–H bonds. The bands at 2111.46 cm⁻1 and 1636.04 cm⁻1 corresponds to the –C≡C– and C–C stretching of aromatic ring. (Fig.4)

Dynamic Light Scattering (DLS)

DLS pattern of the synthesized silver nanoparticles of Goose Berry (*Emblica officinalis*) fruit extract. The size distribution profile indicates that the size (diameter) of these silver nanoparticles varies between 279.04 nm to 2222.51 nm with the average mean size (diameter) of 350.8 nm. (Fig.5)

SEM Analysis

The Scanning Electron Microscope (SEM) micrographs of nanoparticles obtained in the filtrate showed that silver nanoparticles are spherical shaped, well distributed without aggregation. (Fig.6 and Fig.7)

Antibacterial activity

The agar disc diffusion method was employed for screening of antibacterial activities of silver nanoparticles compared with crude extract concentration of $30\mu l$ and $50\mu l$. Vancomycin and Gentamycin is used as a positive control with clear zone of inhibition. The zone of inhibition was observed after 24h incubation at 37°C (Fig 3). The inhibition zone of samples against Gram-positive (*S.aureus*) and Gram-negative (*Klebsiella spp.*) are shown in Table 1 and Table 2. The concentration increased the area of inhibition is increased in both crude and silver nanoparticles. At the low concentration $30\mu l$ and $50\mu l$ of silver nanoparticles has antibacterial better than and Crude extract. (Fig.8 and Table.1)

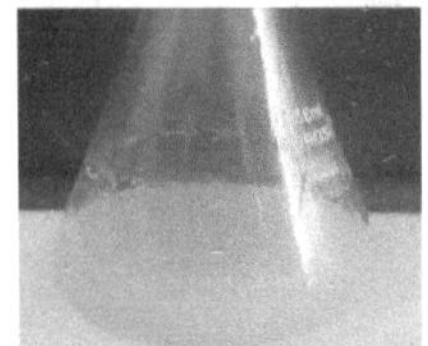

Crude Extract of Wheat Grass

AgNP Extract of Wheat Grass

Crude Extract of Goose Berry

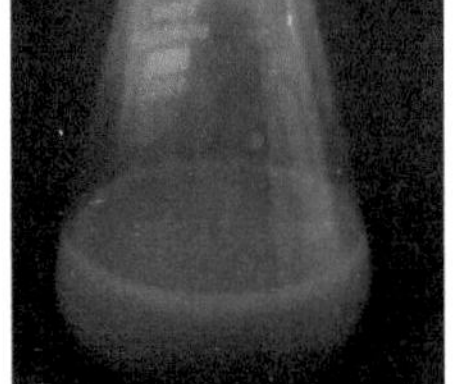

AgNP Extract of Goose Berry

Fig.2 UV-Visible Spectroscopy measurements of *T.aestivum*

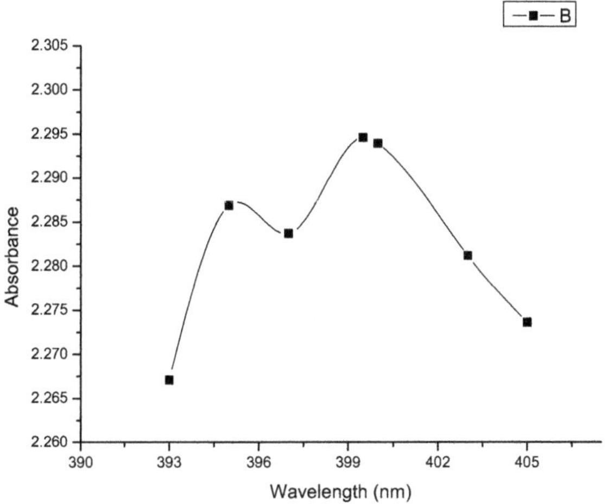

Fig.3 UV-Visible Spectroscopy measurements of *E.officinalis*

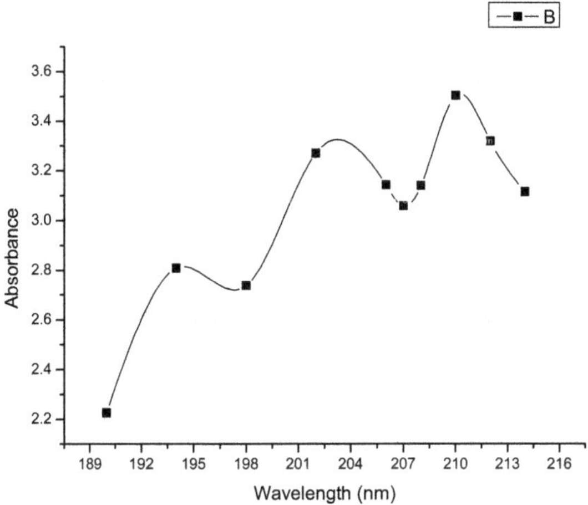

Fig.4 Graphs obtained from FTIR analysis of AgNPs of Wheat Grass Extract.

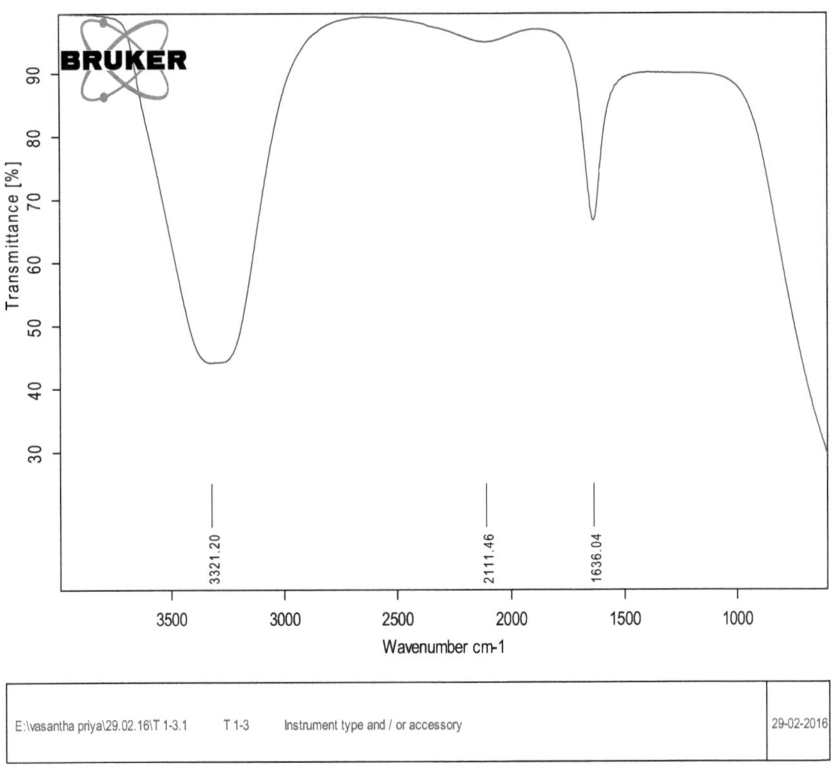

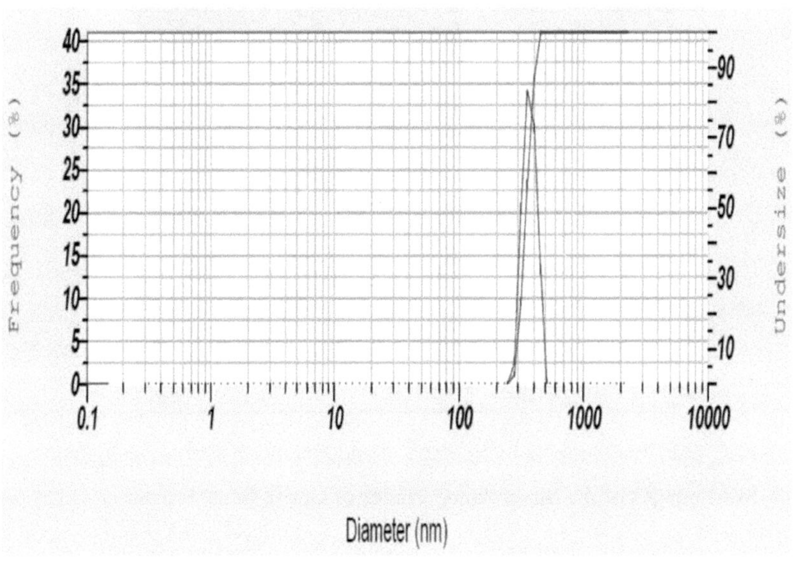

No.	Diameter	Frequency	Cumulation	No.	Diameter	Frequency	Cumulation	No.	Diameter	Frequency	Cumulation
1	0.34	0.000	0.000	26	7.17	0.000	0.000	51	151.57	0.000	0.000
2	0.38	0.000	0.000	27	8.10	0.000	0.000	52	171.25	0.000	0.000
3	0.43	0.000	0.000	28	9.15	0.000	0.000	53	193.48	0.000	0.000
4	0.49	0.000	0.000	29	10.34	0.000	0.000	54	218.60	0.000	0.000
5	0.55	0.000	0.000	30	11.68	0.000	0.000	55	246.98	0.000	0.000
6	0.62	0.000	0.000	31	13.20	0.000	0.000	56	279.04	2.368	2.368
7	0.70	0.000	0.000	32	14.91	0.000	0.000	57	315.27	20.420	22.788
8	0.80	0.000	0.000	33	16.84	0.000	0.000	58	356.20	34.079	56.868
9	0.90	0.000	0.000	34	19.03	0.000	0.000	59	402.44	29.981	86.848
10	1.02	0.000	0.000	35	21.50	0.000	0.000	60	454.69	13.152	100.000
11	1.15	0.000	0.000	36	24.29	0.000	0.000	61	513.71	0.000	100.000
12	1.30	0.000	0.000	37	27.45	0.000	0.000	62	580.41	0.000	100.000
13	1.47	0.000	0.000	38	31.01	0.000	0.000	63	655.76	0.000	100.000
14	1.66	0.000	0.000	39	35.03	0.000	0.000	64	740.89	0.000	100.000
15	1.87	0.000	0.000	40	39.58	0.000	0.000	65	837.07	0.000	100.000
16	2.11	0.000	0.000	41	44.72	0.000	0.000	66	945.74	0.000	100.000
17	2.39	0.000	0.000	42	50.53	0.000	0.000	67	1068.52	0.000	100.000
18	2.70	0.000	0.000	43	57.09	0.000	0.000	68	1207.24	0.000	100.000
19	3.05	0.000	0.000	44	64.50	0.000	0.000	69	1363.97	0.000	100.000
20	3.45	0.000	0.000	45	72.87	0.000	0.000	70	1541.04	0.000	100.000
21	3.89	0.000	0.000	46	82.33	0.000	0.000	71	1741.10	0.000	100.000
22	4.40	0.000	0.000	47	93.02	0.000	0.000	72	1967.14	0.000	100.000
23	4.97	0.000	0.000	48	105.10	0.000	0.000	73	2222.51	0.000	100.000
24	5.61	0.000	0.000	49	118.74	0.000	0.000				
25	6.34	0.000	0.000	50	134.16	0.000	0.000				

Fig.6 SEM images of the Silver nanoparticles of Wheat Grass Extract

Fig.7 SEM images of the Silver nanoparticles of Goose Berry Extract

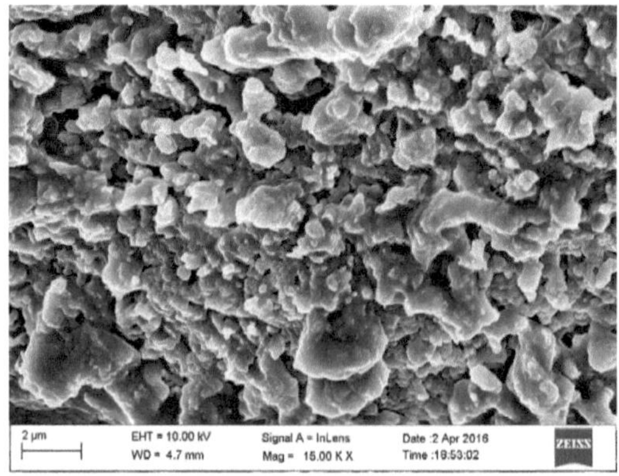

Fig.8 Antibacterial activities of the Wheat Grass and Goose Berry samples

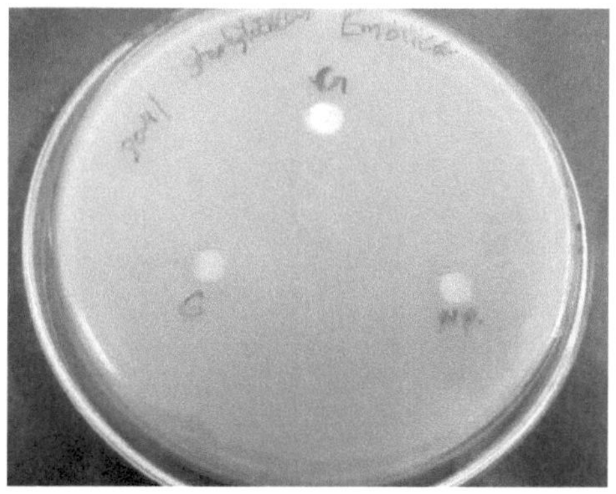

30*µl* Concentration *E.officinalis* in *S.aureus*

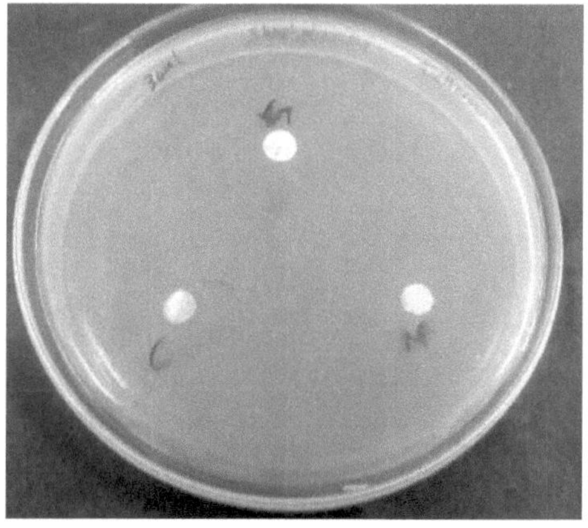

30*µl* Concentration *T.aestivum* in *S.aureus*

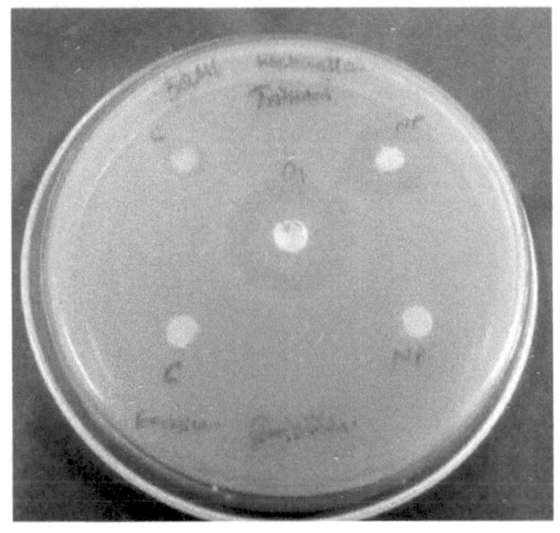

50μl Concentration of *Klebsiella spp.* in Wheat Grass and Goose Berry

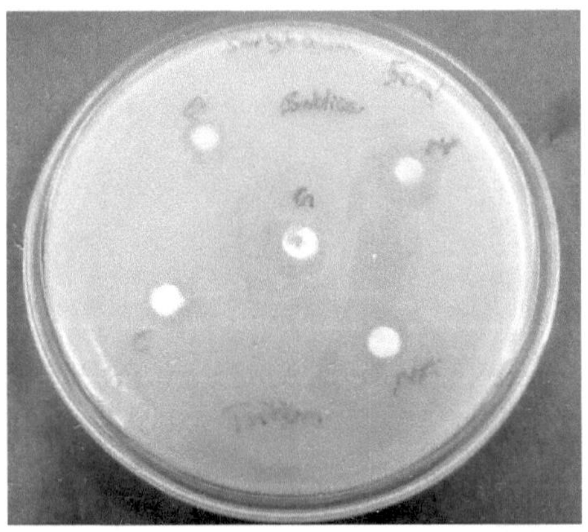

50μl Concentration of *S.aureus* in Wheat Grass and Goose Berry

Table.1 Zone of Inhibition rate of Antimicrobial activity

	30μl		50μl l	
Staphylococcus aureus	Wheat Grass	Goose Berry	Wheat Grass	Goose Berry
Crude Extract	-	±6mm	±9mm	±8mm
Silver Nanoparticles	-	±7mm	±10mm	±12mm
Gentamycin	-	±7mm	±10mm	±10mm
Klebsiella spp.	Wheat Grass	Goose Berry	Wheat Grass	Goose Berry
Crude Extract	±3mm	±5mm	±6mm	±8mm
Silver Nanoparticles	±8mm	±5mm	±15mm	±9mm
Vancomycin	±6mm	±8mm	±13mm	±13mm

Values are Zone of Inhibition ± standard deviation of two replicates

DISCUSSION

Synthesis of silver nanoparticles in plants has been already reported by several authors in the literature. In our present study we have attempted the synthesis of silver nanoparticles of 1mM and 20mM concentration of Wheat Grass and Goose berry extract. The time taken for the synthesis of the particles was minimum 12 hours.

Silver nanoparticles of 1mm concentration were synthesized by using Leaves of Wheat Grass extract and 20mM concentration were synthesized by using Fruit of Goose berry extract. The time was taken for synthesis of silver nitrate to silver nanoparticles was 12-24 hours for both of these extract.

Priya banerjee, (2014) has reported after 5 minutes of conversion process silver nanoparticles showed reddish brown to colloidal brown, suggested the formation of silver nanoparticles in solution in their results.

Shakeel Ahmed and Saiqa Ikram,(2015) has reported after 15 minutes of conversion process silver nanoparticles showed yellow to dark-yellow or dark-brown, suggested the formation of silver nanoparticles in solution in their result.

Kaushik Roy *et al.*, (2014) has reported absorption peak at 460nm for the Parsley Leaves which is differs to the result we obtained for silver nanoparticles of Wheat Grass and Goose Berry extract.

In our present study the FTIR characteristic peak for *T.aestivum* was observed at 3321.20 cm⁻1 indicating the presence of Alkyne group and other bands represents the presence of Aromatic compound which would have played a role in synthesis of AgNPs and they acts as a capping agent for the synthesized silver nanoparticles.

Kaushik Roy, (2015) has reported FTIR band at 2931 cm⁻1 which has identified as Aldehyde group, which due to stretch of C-H bonds in the Aldehyde group. This result almost supported our result for the *T.aestivum* nanoparticles.

Natthawan Phuphansri, (2012) has reported FTIR band at 3342 cm⁻1 which has identified as Hydroxyl group, which due to stretch of C-H bonds in the Hydroxyl group. This result almost supported our result for the *T.aestivum* nanoparticles.

In our present study the DLS Pattern of the nanoparticles of *E.officinalis* was observed at the range between 279.04 nm to 2222.51 nm with the average mean size (diameter) of 350.8 nm.

Kaushik Roy, (2013) has reported the size range of nanoparticles from the DLS pattern between the 2 to 40 nm with average mean size (diameter) of 19 nm.

Kumaravel Palanisamy, (2014) has reported the images of silver nanoparticles of leaves of *E.officinalis* from the SEM analysis the range of the nanoparticles is 135-595nm and it seems to be spherical in morphology.

In our present study the surface morphology of the silver nanoparticles of Leaves of *T.aestivum* and Fruit of *E.officinalis* has showed in the range of 2μm and 1μm in the size.

In our findings we have observed the Gram negative organism *Klebsiella spp.* and Gram Positive organism *S.aureus* was found more susceptible to the synthesized AgNP of both the *T.aestivum* and *E.officinalis* Extract. Both the extract was found effective active against *S.aureus* and *Klebsiella spp.*

Kumaravel Palanisamy, (2014) in their study has reported that silver nanoparticles act primarily against Gram positive bacteria.

SUMMARY

In the present study the leaves of Wheat Grass and the Fruit of Goose Berry was taken for the study of synthesis of silver nanoparticles. The silver nitrate was carried out and 1mM and 20mM concentration was used for nanoparticles synthesis.

The time taken for synthesis of silver nitrate was 12 – 16 hours for both Wheat Grass and Goose Berry Extract. The Color was changed from Yellowish to Dark-Brown.

After synthesis the peak was determined using UV-Vis spectroscopy and it was found to be in the range of 200nm – 400nm for both the samples in this study.

The synthesized silver nitrate was determined using FTIR study. The results indicates the presence of Alkenes group and Aromatic compounds in the synthesized silver nanoparticles of the Wheat Grass Extract.

The Size range of synthesized silver nanoparticles of *E.officinalis* was determined by using DLS method. The size range of the silver nanoparticles of *E.officinalis* between the 279.04 nm to 2222.51 nm with the average mean size (diameter) of 350.8 nm.

The surface morphology of the Silver nanoparticles of Wheat Grass and Goose berry was obtained using SEM images which shows the nanoparticles range of 2μm and 1μm.

The anti-bacterial activity was tested using disk diffusion method. The zone of inhibition was obtained against *S.aureus* and *Klebsiella spp* with good inhibition rate when compared to crude extracts.

LIST OF PUBLICATIONS

1) **Green Synthesis And Antibacterial Activities Of Silver Nanoparticles**, Modern Biological Sciences, SNMV/IICPT/Symbios/2016/OT-031; 1-147. ISBN: 5-8-978-81-926250-0-3

REFERENCES

Ahmad, R. Pandey, S. Sharma, G.K. Khuller. Alginate nanoparticles as   antituberculosis drug carriers: formulation development, pharmacokinetics and therapeutic potential Ind. J. Chest Dis. Allied Sci., 48 (2005), pp. 171–176.

AlRehaily AJ, Al-Howiriny TA, Al-Sohaibani MO, Rafatullah S. Gastroprotective effects of 'amla' Emblica officinalis on invivo test models in rats. Phyto medicine, 2002;9: 515-522.

Ashok kumar S, Ravi S, Velmurugan S. Green synthesis of silver nanoparticles from Gloriosa superba L. leaf extract and their catalytic activity. Spectrochi Acta A, 2013; 115: 388-392.

B. Wiley, Y. Sun, B. Mayers, Y. Xi Chem.-Eur. J., 11 (2005), p. 454.

Bindhu MR, Umadevi M. Synthesis of monodispersed silver nanoparticles using Hibiscus cannabinus leaf extract and its antimicrobial activity. Spectrochi Acta A, 2013; 101: 184-190.

Chandran,S. P., Minakshi Chaudhary., Renu Pasricha., Absar Ahmad and Murali Sastry. 2006. Synthesis of gold nanotriangles and silver nanoparticles using Aloevera plant extract. Biotechnology Progress, 22:577-583.

Daizy Philip. Mangifera Indica leaf assisted biosynthesis of well-dispersed silver

Das J, Paul Das M, Velusamy P. Sesbania grandiflora leaf extract mediated green

De S, Ravishankar B, Bhavsar GC. Plants with hepatoprotective activity- a review. Indian Drugs, 1993; 30: 355-363.

Dharmananda S. Emblic Myrobalans: Amla, Institute of Traditional Medicine. Ethnopharmacol, 2001; 75: 65-69.

Drexler, K. Eric (1986). Engines of Creation: The Coming Era of Nanotechnology. Doubleday. ISBN 0-385-19973-2.

Drexler, K. Eric (1992). Nanosystems: Molecular Machinery, Manufacturing, and Computation. New York: John Wiley & Sons. ISBN 0-471-57547-X.

F. Kruis, H. Fissan, B. Rellinghaus Mater. Sci. Eng. B, 69 (2000), p. 329.

F. Mafune, J. Kohno, Y. Takeda, T. Kondow, H. Sawabe J. Phys. Chem. B, 104 (2000), p. 9111.

Gardea Torresdey JL., Parsons GL., Gomez E., Peralta Videa J., Troiani HE., Santiago P., Jose Yacaman M. Formation and Growth of Au nanoparticles inside live *Alfalfa* Plant. Nano Lett, 2002; 2(4): 397-401.

Goodshell DS (2004) Bionanotechnology. Lessons from Nature. John Wiley & Sons Inc. Publication.

H. Gu, P.L. Ho, E. Tong, L. Wang, B. Xu Presenting vancomycin on nanoparticles to enhance antimicrobial activities Nano Lett., 3 (9) (2003), pp. 1261–1263.

H. Huang, Y. Yang. Compos. Sci. Technol., 68 (2008), p. 2948.

J. Köhler, L. Abahmane, J. Albert, G. Mayer. Chem. Eng. Sci., 63 (2008), p. 5048.

J.L.G. Torresdey, E. Gomez, J.R.P. Videa, J.G. Parsons, H. Troiani, M.J. Yacaman *Alfalfa* sprouts: a natural source for the synthesis of silver nanoparticles Langmuir, 19 (2003), pp. 1357–1361.

Jain D., H.K. Daima., S. Kachhwaha and S.L. Kothari, 2009. Synthesis of plant-mediated silver nanoparticles using papaya fruit extract and evaluation of their anti-microbial activities, Digest Journal of nanomaterials and Biostructure, 4: 557-563.

Jeena KJ, Kuttan R. Antioxidant activity of Emblica officinalis. J Clin Biochem Nutr, 1995; 19: 63-70.

Jose JK, Kuttan Y, Kuttan R. Antitumour activity of Emblica officinalis. J K. Kelly, E. Coronado, L. Zhao, G. Schatz J. Phys. Chem. B, 107 (2003), p. 668.

Kaushik Roy., Supratim Biswas and Pataki C Banerjee. 'Green' Synthesis of Silver Nanoparticles by Using Grape (*Vitis venifera*) Fruit Extract: Charaterization of the particles and study of Antibacterial activity. RJPBCS, 2013; 4(1): 0975-8585.

Kumaravel Palanisamy., Prabhakaran Arumugam Kalaiselvi., Melchias Gabriel., Jeyanthi Thangavel., Logesh Sundaram. *Emblica officinalis* Extract mediated Green synthesis of Antibacterial silver nanoparticles against Human Pathogens. Wjpps, 2014; 3(3): 2278-4357.

Krutyakov Y, A. Olenin, A. Kudrinskii, P. Dzhurik, G. Lisichkin Nanotechnol. Russia, 3 (5–6) (2008), p. 303.

L. Li, J. Hu, A.P. Alivistos. Nano Lett., 1 (2001), p. 349.

M. Oliveira, D. Ugarte, D. Zanchet, A. Zarbin J. Colloid Interface Sci., 292 (2005), p. 429.

M.A. Albrecht, C.W. Evans, C.L. Raston Green chemistry and the health implications of nanoparticles Green Chem., 8 (2006), pp. 417–432.

Melina, Vesanto, MS, RD & Davis, Brenda, RD: "The New Becoming Vegetarian", page 186-187. Healthy Living Publications, 2003.

Mervat F Zayeda, Wael H Eisa, Shabaka AA. Malva parviflora extract assisted green synthesis of silver nanoparticles. Spectrochi Acta A, 2012; 98: 423-428.

Meyerowitz, Steve (April 1999). "Nutrition in Grass". Wheatgrass Nature's Finest Medicine: The Complete Guide to Using Grass Foods & Juices to Revitalize Your Health (6th ed.). Book Publishing Company. p. 53. ISBN 1-878736-97-3.

Mittal A.K., Chisti Y & Banerjee U.C. (2013). Synthesis of metallic nanoparticles using plant extracts. *Biotechnology Advances*, 31, 346-356.Nanoparticles using aqueous extract and dried leaf of *Anacardium occidentale*. Spectrochi Acta A, 2011; 279: 254-262. nanoparticles. Spectrochi Acta A, 2011; 78: 327-331.

P. Gong, H. Li, X. He, K. Wang, J. Hu, W. Tan. Preparation and antibacterial activity of $Fe_3O_4@Ag$ nanoparticles Nanotechnology, 18 (2007), pp. 604–611.

P.S. Retchkiman-Schabes, G. Canizal, R. Becerra-Herrera, C. Zorrilla, H.B. Liu, J.A.Ascencio Biosynthesis and characterization of Ti/Ni bimetallic nanoparticles Opt. Mater., 29 (2006), pp. 95–99.

Perianayaham JB, Narayanan G, Gnanasekar G, Pandurangan S, Raja S, Rajagopal K. Evaluation of antidirrheal potential of *Emblica officinalis*. Pharm Biol, 2005; 43: 373-377.

Prashant Mohanpuria., Nisha K. Rana & Sudesh Kumar Yadav. Biosynthesis of nanoparticles: technological concepts and future applications. J Nanopart Res (2008) 10:507–517.

Raja K, Saravanakumar A, Vijayakumar R. Efficient synthesis of silver nanoparticles from Prosopis juliflora leaf extract and its antimicrobial activity using sewage. Spectrochi Acta A, 2012; 97: 490-494.

Rupali S Patil, Mangesh R Kokate, Sanjay S Kolekar. Bioinspired synthesis of highly stabilized silver nanoparticles using Ocimum tenuiflorum leaf extract and their antibacterial activity. Spectrochi Acta A, 2012; 91: 234-238.

Kim S, B. Yoo, K. Chun, W. Kang, J. Choo, M. Gong, S. Joo J. Mol. Catal. A: Chem., 226 (2005), p. 231.

S.S. Shankar, A. Ahmad, M. Sastry Geranium leaf assisted biosynthesis of silver nanoparticles Biotechnol. Prog., 19 (2003), pp. 1627–1631.

S.S. Shankar, A. Ahmad, R. Pasricha, M. Sastry. Bioreduction of chloroaurate ions by geranium leaves and its endophytic fungus yields gold nanoparticles of different shapes J. Mater. Chem., 13 (2003), pp. 1822–1826.

Sabu MC, Kuttan R. Antidiabetic activity of medicinal plants and its relationship with their antioxidant property. J Ethnopharmacol, 2002; 81: 155-160.

Sai Ram K, Ram CV, Dora Babu M, Vijay Kumar K, Agrawal VK, Goel K.Antiulcerogenic effect of methanolic extract of Emblica officinalis: an experimental study. J Ethnopharmacol, 2002; 821-829.

Samhita C. Ed., translation by the Shree Gulabkunverba Society, Volume 4. Chikitsa Sthana, Jamnagar, India: 1949.

Seymour, Kent. "WHEAT GRASS (Triticum aestivum)" (PDF). Illinois State University. Retrieved 11 December 2013.

Sharma V., Shukla R.K., Saxena N., Parmar D., Das M & Dhawan A. (2009). DNA damaging potential of zinc oxide nanoparticles in human epidermal cells. Toxicology Letters, 185, 211-218.

Sheny DS, Joseph M, Daizy P. Phytosynthesis of Au, Ag and Au–Ag bimetallic

Shikha Behera and P.L. Nayak. Green synthesis and characterization of Zero Valent Silver Nanoparticles from the Extract of *Vitis venifera*. World J. Nano Sci. Technol., 2(1): 58-61, 2013.

Shoba G., Vinutha Moses & Ananda S. Biological Synthesis of Copper nanoparticles and its impact – a Review. IJPSI, 2014; 8(3): 2319-6718.

Stephen JR., Macnaughton SJ. Development in terrestrial bacterial remediation of metals. Curr Opin Biotechnol, 1999; 10(3): 230-233.

Synthesis of antibacterial silver nanoparticles against selected human pathogens. Spectrochi Acta A, 2013; 104: 265-270.

T. Tsuji, K. Iryo, N. Watanabe, M. Tsuji Appl. Surf. Sci., 202 (2002), p. 80.

T. Tsuji, T. Kakita, M. Tsuji Appl. Surf. Sci., 206 (2003), p. 314.

Thakur CP, Thakur B, Singh B, Singh S, Sinha PK, Sinha SK. The Ayurvedic medicines, Haritaki, Amla and Bahira reduce cholesterol induced atherosclerosis in rabbits. Indian J Cardiol, 1998; 21: 167-175.

Thiruppathi C, Kumaravel P, Duraisamy R, Prabhakaran AK, Jeyanthi T, Sivaperumal R, Karthick PA. Biofabrication of Silver nanoparticles using Cocculus hirsutus leaf extract and their antimicrobial efficacy. Asian J Pharm Tech, 2013; 3: 93-97.